TOXICITY

OF

CHEMICALS ON FISH

Explore the impact of chemicals
on fish focusing on acute toxicity
and it effects on aquatic
ecosystem.

By

SEBASTIAN WAYNE

Table of content

Introduction

The effect of chemicals on fish is an important issue in the field of aquatic ecosystems. Acute toxicity refers to the rapidly occurring adverse effects of a chemicals within a short period of time (usually within 96 hours after exposure). Understanding the toxicity of chemicals on fish is important for assessing environmental risks, protecting biodiversity and ensuring the sustainability of aquatic life. In this E-book we explore the complex interactions between chemicals and fish, exploring mechanisms of acute toxicity, factors affecting susceptibility, and methods for assessing toxicity levels. By presenting these complex issues, we aim to provide readers with the information needed to monitor and reduce problems caused by chemicals in water bodies.

A brief description of this book

It will address the issue of toxicity in fish related to short-term (usually 96 hours or less) exposure to chemicals caused by the product quickly and painfully. Toxic chemicals such as heavy metals and pesticides can affect fish behavior and growth. The severity of toxicity is affected by many factors, including the concentration of chemicals, duration of exposure, fish species, life stage, and environmental conditions. Fish show different responses to acute toxicity; some species are more sensitive than others. Toxicity tests require tests such as standard toxicity tests in which fish are exposed to increasing concentrations of Chemicals to determine lethal and sub-lethal effects. These assessments are important for management purposes, environmental risk assessment and the protection of aquatic biodiversity. Understanding the acute toxicity of chemicals on fish is important for environmental management decisions, pollution control, and watershed protection. By understanding and reducing the risks posed by chemical water, we can work towards a healthy and sustainable water supply.

Chapter one

Understanding chemicals and fish behaviors

Toxicity tests are used to evaluate risks to fish and other health regulations regarding aquatic contamination. Tests were conducted on three different fish species: cold water species, warm water species, and marine/estuarine species. This test is usually done for all acute Toxicity Testing and approximately 260 fish are needed. Fish Acute Toxicity Test (Test No. 203) is a method used to determine the acute toxicity of chemicals on fish. It is a standardized test method designed to assess the acute toxicity of chemicals or substances to fish. This test is commonly used in ecotoxicology and regulatory toxicology to evaluate the potential adverse effects of substances on aquatic organisms. The results of the test can provide valuable information for environmental risk assessments, chemical safety evaluations, and regulatory decision-making processes.

In the Fish Acute Toxicity Test, commonly used fish species, such as rainbow trout (Oncorhynchus mykiss) or zebrafish (Danio rerio), are typically exposed to varying concentrations of the test substance under controlled laboratory conditions. The primary endpoint of the test is the determination of the median lethal concentration (LC50) or median lethal dose (LD50), which represents the concentration or dose of the test substance that is estimated to be lethal to 50% of the exposed fish within a specified period, usually 96 hours for acute tests.

The test follows standardized protocols, such as those outlined by organizations like the Organization for Economic Co-operation and Development (OECD) or the US Environmental Protection Agency (EPA), to ensure consistency and reproducibility of results across different testing laboratories. These protocols provide detailed guidance on test design, fish species selection, test substance preparation, exposure conditions, observation of test organisms, and data analysis.

During the test, fish are typically exposed to a range of concentrations of the test substance and a control group without the test substance. The exposure concentrations are often chosen based on prior knowledge of the substance and its expected toxicity, and they may include geometric series of dilutions to cover a broad range of potential effects. The fish are observed at regular intervals to record mortality, abnormal behavior, and other signs of toxicity. These observations are used to determine the concentration-response relationship and to calculate the LC50 or LD50 using statistical methods.

The Fish Acute Toxicity Test can also include additional endpoints, such as sub lethal effects (e.g., changes in fish behavior, growth, or reproductive success), histopathological examinations, or biochemical analyses to provide a more comprehensive understanding of the toxic effects of the test substance on fish.

The results of the test provide important information for assessing the potential environmental impact of chemicals, especially in aquatic ecosystems. Regulatory agencies may use the data from Fish Acute Toxicity Tests to determine safe concentrations for the discharge of substances into water bodies, to set environmental quality standards, or to make decisions about the approval, restriction, or banning of chemicals based on their potential risks to aquatic life.

It's important to note that the Fish Acute Toxicity Test is just one component of a comprehensive

Eco-toxicological assessment, which may also include tests with other aquatic organisms (e.g., invertebrates, algae), as well as studies on the fate and behavior of the test substance in the environment. When interpreting the results of the test, factors such as the potential for bioaccumulation, persistence, and ecological interactions should also be considered to fully understand the environmental implications of the test substance. Toxicity tests provide rapid and reproducible data that can be used to determine relative concentrations of toxic chemicals for different fish species. These tests also help determine the significance and design of additional toxicity studies. Toxicity test results are submitted to the EPA Toxicology Reference Library. In an effort to reduce the use of animals in fish toxicity testing, the ICCVAM Ecotoxicology Working Group has published a summary of the requirements for eco-toxicity testing of selected federal agencies involved in eco-toxicity assessment and environmental safety. And potential applications of these technologies for regulatory evaluation. To better understand the toxicity of chemicals on fish, it is important to consider their effects on fish and ecosystems. Toxicity tests provide information about the lethality of a chemical in fish but do not imply sub-lethal effects such as changes in behavior, growth, or development. Sub-lethal effects can have significant impacts on fish and ecosystems and may not be detected by toxicity tests alone. In addition to toxicity tests, other toxicity tests may be used to evaluate the potential effects of chemicals on fish and other aquatic organisms. For example, chronic toxicity testing can provide information about the long-term effects of exposure to low doses of chemicals. Bioassays use living organisms to investigate the presence and effects of toxic substances in water or sediment samples and can also be used to evaluate the chemical potential of fish and other aquatic organisms. It is important to consider the potential of chemicals in aquatic ecosystems. Chemicals can enter water through a variety of pathways, including industrial and urban emissions, agricultural runoff, and atmospheric deposition.

Understanding the sources and pathways of chemical contamination can help identify potential impacts on fish and ecosystems and guide strategies to reduce or prevent contamination. Overall, understanding the toxicity of chemicals to fish is an important part of assessing the risks of chemicals in aquatic ecosystems. However, it is important to consider the potential for sub-lethal complications and the location of the chemical in other types of chemical testing to understand its impact on fish and ecosystems. It is important to consider the toxicity of chemicals to fish and their consequences for human health. Fish is an important source of protein for many people around the world, and exposure to toxic chemicals from contaminated fish can be hazardous to human health. Some chemicals, such as mercury and polychlorinated biphenyls, can accumulate in fish flesh and bio amplify in food, causing more in larger fish. In addition to toxicity tests and other toxicity tests, bio-monitoring can also be used to evaluate the potential effects of toxic chemicals on fish and human health. Bio-monitoring involves measuring the concentration of chemicals in fish flesh and can provide information about contamination and potential risks to human health. Regulatory agencies, such as the U.S. Environmental Protection Agency (EPA), use a variety of methods to evaluate the risks of toxic chemicals to fish and human health. These procedures include setting water quality standards, establishing fish consumption rules, and regulating the release of chemicals into the water. Overall, understanding the toxicity of chemicals to fish is an important part of assessing the risks of chemicals in aquatic ecosystems.

However, to understand the potential impact on fish, ecosystems and human health, it is important to consider possible impacts on human health, as well as other biological tests, bio-monitoring and control methods. To better understand the toxicity of chemicals on fish, it is important to consider their possible effects on human health, Fish is an important source of protein for many people around the world, and exposure to toxic chemicals from contaminated fish can be hazardous to human health. Some chemicals, such as mercury and PCBs, can accumulate in fish flesh and bio-amplify in food products, causing more in larger fish. In addition to clinical toxicity tests and other toxicity tests, bio-monitoring may be used to evaluate the potential effects of toxic chemicals on fish and human health. Bio-monitoring involves measuring the concentration of chemicals in fish flesh and can proved information about contamination and potential risks to human health. Regulatory agencies, such as the U.S. Environmental Protection Agency (EPA), use a variety of methods to evaluate the risks of chemicals to fish and human health, These procedures include setting water quality standards, establishing fish consumption rules, and regulating the release of chemicals into the water.

Impact of Fertilizer chemicals on Fish

Fertilizers are used in agriculture to enhance plant growth by providing essential nutrients such as nitrogen, phosphorus, and potassium. However, when these fertilizers are not managed properly, they can have adverse effects on fish and aquatic life.

One of the primary concerns regarding the impact of fertilizers on fish is nutrient pollution. When excess fertilizers are applied to fields, the nutrients can runoff into nearby water bodies during rain events. This runoff carries high levels of nitrogen and phosphorus, which can lead to eutrophication of the water. Eutrophication occurs when excessive nutrient levels stimulate the growth of algae and aquatic plants. While this may seem beneficial at first, the subsequent decay of these organisms can deplete oxygen levels in the water, leading to hypoxic or anoxic conditions. This can be devastating for fish and other aquatic organisms, as it can result in fish kills and the loss of important habitat.

In addition to eutrophication, the presence of excess nutrients in water bodies can also lead to harmful algal blooms. Certain species of algae produce toxins that are harmful to fish and other aquatic organisms. These toxins can cause fish kills and have the potential to bioaccumulate in the food chain, posing a risk to human health if contaminated fish are consumed.

Furthermore, high levels of nutrients in water bodies can alter the natural balance of aquatic ecosystems. Excessive plant growth fueled by nutrient pollution can choke waterways, reducing water flow and hindering the movement of fish. It can also lead to changes in the composition of aquatic plant and animal communities, potentially favoring invasive species over native ones. This can have cascading effects throughout the ecosystem, impacting the entire food web and leading to a decline in fish populations.

Fertilizers can also indirectly impact fish by contributing to the degradation of their habitat. Nutrient pollution can lead to sedimentation, as excessive plant growth and algal blooms trap sediments and organic matter, reducing water clarity and smothering important fish spawning grounds and nursery areas. Sedimentation can also clog fish gills, suffocate fish eggs, and smother benthic invertebrates, which are important food sources for fish.

The impact of fertilizers on fish is not limited to freshwater environments. In fact, nutrient pollution from fertilizers can have far-reaching effects in coastal and marine ecosystems as well. Excess nutrients can lead to the degradation of estuaries and coral reefs, impacting fish populations that rely on these critical habitats for feeding, spawning, and shelter.

It is important to note that the impact of fertilizers on fish is not solely a result of agricultural practices. Urban and suburban areas also contribute to nutrient pollution through the use of lawn and garden fertilizers.

Additionally, industrial activities, such as poorly managed aquaculture operations and wastewater discharges, can also be significant sources of nutrient pollution in water bodies, further exacerbating the impact on fish.

Efforts to mitigate the impact of fertilizers on fish and aquatic environments are underway. Best management practices in agriculture aim to reduce nutrient runoff through measures such as precision nutrient application, cover crops, buffer strips, and conservation tillage. These practices help to keep nutrients in the soil and out of water bodies, ultimately reducing the potential for eutrophication and its associated impacts on fish.

In urban and suburban areas, responsible fertilizer use and landscaping practices can play a role in minimizing nutrient pollution. This includes selecting fertilizers with slow-release formulations, following proper application rates and timing, and utilizing native plants that require fewer inputs and have lower nutrient needs. Proper maintenance of septic systems and the use of advanced wastewater treatment technologies also help to reduce nutrient loading in water bodies.

Regulatory measures, such as the enforcement of nutrient management regulations and the establishment of water quality standards, are essential for addressing the impact of fertilizers on fish. These measures can help to minimize nutrient pollution from various sources and promote sustainable practices that protect fish and aquatic ecosystems.

Public education and outreach efforts are also crucial in raising awareness about the impact of fertilizers on fish and empowering individuals to take action in their own communities. By promoting practices that reduce nutrient runoff, such as responsible fertilizer use, proper disposal of pet waste, and the maintenance of riparian buffers, communities can contribute to the protection of fish habitat and water quality.

The effect of chemical fertilizers on fisheries When it comes to pollution, it is noted that the most important chemical pollution in the oceans is ecological damage, Additionally, agricultural chemical fertilizers have been shown to be harmful to aquatic biota. Studies have shown that with certain combinations of fertilizers in the aquatic environment, algae and aquatic plants grow uncontrollably and clog waterways, resulting in depletion of dissolved oxygen in the system. Note that excessive use of inorganic fertilizers can affect the structure and function of water in the community and decrease oxygen in the water. Other researchers have shown that the interaction between Chemicals and fish has a significant effect on tail beating and ventilation, and bleeding of skin, gill filament, and fins it can result to dead of fish. NPK fertilizers have also been reported to have sub-lethal effects on fish, This results in poor growth and death due to the toxic chemicals used in the treatment. Phosphorus-potassium fertilization on fish can cause poor growth and death due to the presence of toxic chemicals used in the treatment.

Fish Behavior

Fish behavior is a fascinating and diverse field of study that encompasses a wide range of behaviors exhibited by fish species in various aquatic environments. Understanding fish behavior is crucial for ecologists, fisheries biologists, aqua-culturists, and conservationists as it provides insights into the ecological roles of fish, their interactions with other species, and their responses to environmental changes. From feeding and reproduction to migration and communication, fish behavior is influenced by a myriad of factors including genetics, physiology, habitat, and social interactions.

Feeding behavior is a fundamental aspect of fish behavior that greatly influences their survival and ecosystem dynamics. Fish exhibit a wide array of feeding strategies, from filter feeding and grazing to predation and scavenging, depending on their species and ecological niche. For example, pelagic fish such as herring and anchovies engage in filter feeding, where they swim with their mouths open to capture plankton, while predatory fish like barracuda and sharks actively hunt and capture prey using specialized feeding adaptations such as sharp teeth and powerful jaws.

Reproductive behavior is another critical aspect of fish behavior, with species displaying diverse strategies for courtship, mate selection, and parental care. Some fish species engage in elaborate courtship rituals involving displays of color, fin movements, and sound production to attract mates and establish breeding territories. In some cases, males compete for access to females, while females choose and evaluate potential mates based on various cues such as coloration, body size, and behavior. After fertilization, some fish species exhibit different forms of parental care, ranging from no care at all to complex care behaviors such as nest building, egg guarding, and fry protection.

Migration is a remarkable behavior exhibited by many fish species and plays a crucial role in their life cycles and population dynamics. Fish migration can encompass a wide range of movements, including daily foraging migrations, seasonal movements for spawning, and long-distance oceanic migrations. For example, salmon undertake extensive spawning migrations from the ocean to freshwater rivers, navigating complex and often hazardous environments to reach their natal spawning grounds. Eels are known for their remarkable catadromous migration, where they travel from freshwater rivers to the ocean to spawn. Understanding migration patterns is essential for effective fisheries management and conservation efforts, as disruptions to migration routes can have significant impacts on fish populations.

Social behavior in fish is another intriguing aspect of their behavior, as many species exhibit complex social structures and interactions. Fish engage in various forms of social behavior, including shoaling, schooling, territoriality, and hierarchical structures.

Shoaling and schooling behavior provide fish with benefits such as predator avoidance, foraging efficiency, and communication. Territorial behavior is exhibited by many fish species, where individuals defend specific areas against intruders to secure resources or breeding sites. In some species, individuals establish social hierarchies based on size, age, or dominance, which can influence access to food, mates, and shelter.

Communication in fish involves a diverse array of visual, tactile, chemical, and acoustic signals used for navigation, foraging, predator avoidance, courtship, and social interactions. Visual signals, such as body coloration and fin movements, play a crucial role in courtship displays and aggressive interactions. Tactile communication involves physical interactions such as fin displays and body contact, which can convey information about social status and reproductive readiness. Chemical communication, through the release of pheromones, allows fish to convey information about territory boundaries, reproductive status, and alarm cues. Acoustic communication, in the form of vocalizations and sound production, is prevalent in many fish species and serves various functions such as attracting mates, establishing territory, and coordinating group movements.

Environmental factors play a significant role in shaping fish behavior, as fish exhibit remarkable adaptability in response to changes in their habitat. Factors such as water temperature, oxygen levels, salinity, water clarity, and habitat structure influence the distribution, movement, and behavior of fish species. For example, changes in water temperature can influence the timing of spawning events and the migratory patterns of fish. Similarly, alterations in habitat structure due to human activities can impact fish behavior, leading to changes in foraging habits, shelter selection, and social interactions.

In conclusion, fish behavior encompasses a rich tapestry of actions and interactions that are shaped by evolutionary, ecological, and environmental factors. Studying fish behavior provides valuable insights into the functioning of aquatic ecosystems, the management of fisheries resources, the conservation of endangered species, and the welfare of aquaculture species. By gaining a deeper understanding of fish behavior, we can work towards promoting the sustainable management of fish populations, the protection of critical habitats, and the preservation of the intricate and diverse behaviors exhibited by these remarkable aquatic creatures.

Understanding the acute toxicity of chemicals in fish

Understanding the acute toxicity of chemicals on fish is an essential aspect of environmental toxicology and aquatic ecosystem health. Acute toxicity refers to the adverse effects caused by a single or short-term exposure to a toxic substance, resulting in immediate or rapid impacts on the health and survival of fish. Assessing the acute toxicity of chemicals on fish involves understanding the interactions between toxicants and fish at molecular, physiological, and ecological levels, as well as evaluating the potential risks to aquatic ecosystems and human activities such as fisheries and aquaculture.

The assessment of acute toxicity typically involves determining the lethal concentration of a toxicant that causes mortality in a specified percentage of test organisms within a given exposure period. This is expressed as LC50, the concentration of a substance that is lethal to 50% of the test organisms. Additionally, the measurement of sub-lethal effects, such as changes in behavior, physiology, and biochemistry, provides a more comprehensive understanding of the impacts of chemical exposure on fish.

The acute toxicity of chemicals on fish is influenced by various factors, including the chemical properties of the toxicant, the physiological characteristics of the fish species, and the environmental conditions in which the exposure occurs. Water-solubility, chemical stability, and bioavailability are important properties that determine the fate and transport of toxicants in aquatic environments. The mode of action of the toxicant, such as interference with metabolic processes, disruption of ion regulation, or damage to specific organs or tissues, also affects the severity of acute toxicity in fish.

Fish species exhibit varying degrees of sensitivity to different toxicants, with factors such as body size, life stage, and physiological adaptations influencing their susceptibility to acute toxicity. For example, developing embryos and larvae may be more sensitive to chemical exposure due to their underdeveloped detoxification mechanisms and higher surface area-to-volume ratio, which increases their uptake of toxicants from the surrounding water. Similarly, species-specific differences in detoxification enzymes, metabolic rates, and tolerance to environmental stressors contribute to the variability in acute toxicity among fish species.

Environmental conditions, such as water temperature, pH, oxygen levels, and salinity, play a significant role in modulating the acute toxicity of chemicals on fish. Changes in these parameters can affect the bioavailability and uptake of toxicants, as well as the physiological responses of fish to chemical stress. For example, elevated water temperatures can enhance the uptake and toxic effects of certain chemicals on fish, while low oxygen levels can exacerbate the impacts of chemical exposure by compromising the respiratory and metabolic functions of fish.

Assessing the acute toxicity of chemicals on fish involves conducting standardized laboratory tests, such as the acute toxicity tests with fish, which expose test organisms to varying concentrations of toxicants under controlled conditions to determine their lethal and sub-lethal effects. These tests provide valuable data on the dose-response relationships, the time course of toxicity, and the potential impacts on different organs and tissues in fish. In addition to laboratory tests, field studies and monitoring programs are employed to assess the acute toxicity of chemicals in natural aquatic environments, taking into account the complexities of real-world exposure scenarios and the interactions with other environmental stressors.

The evaluation of acute toxicity data allows for the determination of water quality criteria and regulatory standards to protect aquatic organisms from the adverse effects of chemical exposure. Regulatory agencies and environmental management authorities utilize acute toxicity information to establish permissible levels of toxicants in water bodies, develop guidelines for the safe use and disposal of chemicals, and implement measures to prevent and mitigate acute toxicity incidents in aquatic ecosystems.

Understanding the acute toxicity of chemicals on fish is crucial for safeguarding the health and integrity of aquatic ecosystems, as well as for ensuring the sustainable management of fisheries and aquaculture activities. By integrating knowledge of chemical toxicity, fish physiology, and environmental factors, scientists and environmental managers can assess the risks posed by toxicants, implement effective mitigation strategies, and promote the conservation and restoration of healthy aquatic habitats.

The benefits of understanding the acute toxicity of chemicals on fish

Understanding the acute toxicity of chemicals on fish offers a range of benefits that are integral to environmental protection, fisheries

61

management, and human well-being. These benefits extend across various sectors and are crucial for ensuring the sustainable use and conservation of aquatic ecosystems. Here are several key benefits of understanding acute toxicity of chemicals on fish:

1. Environmental Protection: By comprehensively understanding the acute toxicity of chemicals on fish, we can better assess and mitigate the risks posed by various toxicants to aquatic ecosystems. This knowledge enables the development of targeted strategies to minimize pollution, protect

sensitive habitats, and preserve biodiversity. It allows for informed decision-making in environmental management, pollution control, and conservation efforts, leading to the safeguarding of aquatic environments for current and future generations.

2. Human Health and Safety: Understanding acute toxicity in fish provides valuable insights into potential risks to human health associated with the consumption of contaminated fish. By identifying harmful chemicals and their impacts on fish populations, we can better regulate fishing and

aquaculture practices, set safe levels for chemical exposure, and protect human communities that rely on fish as a food source. This contributes to the prevention of health hazards and the promotion of safe and sustainable fisheries.

3. Regulatory Standards and Guidelines: Acute toxicity data are essential for establishing water quality criteria, setting regulatory standards, and developing guidelines for the safe use and disposal of chemicals. This supports the enforcement of environmental regulations, the implementation of best practices

in chemical management, and the promotion of responsible use of substances that could potentially harm aquatic organisms. Regulatory measures driven by a thorough understanding of acute toxicity contribute to the protection and improvement of water quality and aquatic ecosystem health.

4. Sustainable Fisheries and Aquaculture: Knowledge of acute toxicity in fish guides the sustainable management of fisheries and aquaculture activities. By understanding the effects of chemicals on fish

populations, we can minimize the impacts of contaminants, ensure the sustainability of fish stocks, and support the responsible development of aquaculture practices. This benefits both the economic livelihoods of fisheries-dependent communities and the long-term health of aquatic ecosystems.

5. Pollution Prevention and Remediation: Understanding acute toxicity provides the basis for developing proactive measures to prevent pollution and remediate contaminated sites. By identifying the sources and impacts of

toxicants on fish, we can implement strategies to reduce chemical discharges, improve wastewater treatment, and remediate polluted areas. This supports the restoration of degraded aquatic habitats and the reduction of environmental harm from chemical pollutants.

6. Risk Assessment and Mitigation: Acute toxicity data are instrumental in assessing the risks associated with chemical exposure in aquatic environments. This information allows for the identification of high-risk chemicals, vulnerable fish species,

and critical exposure scenarios, enabling the implementation of targeted risk mitigation measures. By understanding acute toxicity, we can develop and implement effective risk management strategies, reducing the potential for adverse impacts on fish and their habitats.

7. Scientific Research and Innovation: Understanding acute toxicity in fish fosters scientific research and innovation in the fields of environmental toxicology, aquatic ecology, and conservation biology. This knowledge fuels advances in the

development of novel toxicity testing methods, the investigation of emerging contaminants, and the exploration of ecological interactions. It also supports the discovery of new approaches for mitigating chemical toxicity and improving the resilience of aquatic ecosystems in the face of environmental stressors.

Chapter Two
Key Concepts and Principles

The concept of acute toxicity of chemicals on fish refers to the adverse effects caused by a single or short-term exposure to high concentrations of toxic substances.

Understanding the principles of acute toxicity is fundamental to assessing the immediate and severe impacts of chemicals on fish and other aquatic organisms. The concept and principles of acute toxicity encompass various aspects of toxicological assessment, environmental risk evaluation, and the development of regulatory standards. Here are key concepts and principles related to the acute toxicity of chemicals on fish:

1.Dose-Response Relationship: The principle of dose-response relationship is central to acute toxicity assessments. It involves determining the correlation between the dose (or concentration) of a toxicant and the magnitude of the biological response in fish. This relationship is typically characterized by dose-response curves, which illustrate how the severity of toxic effects changes as the exposure concentration increases. Understanding the dose-response relationship is essential for evaluating the lethal and sub-lethal effects of toxicants on fish and establishing dose thresholds for adverse effects.

2.Lethal Concentration (LC50): The concept of LC50 represents the concentration of a toxicant in water that is lethal to 50% of the exposed fish population within a specified exposure period. LC50 values are essential indicators of acute toxicity and are commonly used to compare the relative toxicity of different chemicals. Assessing LC50 values allows for the classification of chemicals based on their acute toxicity to fish and facilitates the establishment of water quality criteria and regulatory standards.

3. Acute Toxicity Testing: Acute toxicity testing involves exposing fish to varying concentrations of a chemical over a short period to evaluate the lethal and sublethal effects. Standardized acute toxicity tests, such as the OECD 203 Fish Acute Toxicity Test, provide a systematic approach for determining the acute toxicity of chemicals on fish. These tests adhere to specific protocols and endpoints, enabling the generation of reliable toxicity data for regulatory and risk assessment purposes.

4. Species Sensitivity: The principle of species sensitivity acknowledges that different fish species exhibit varying sensitivities to toxicants. Understanding species-specific differences in acute toxicity is crucial for assessing the potential impacts of chemicals on diverse fish populations in aquatic ecosystems. It also informs the selection of appropriate indicator species for toxicity testing and risk assessment, considering the ecological relevance of the fish species present in the environment of concern.

5. Environmental Factors: Acute toxicity assessments consider the influence of environmental factors on the toxicity of chemicals to fish. Factors such as water temperature, pH, dissolved oxygen, and water hardness can modify the bioavailability and toxicity of chemicals in aquatic environments. Understanding the interactions between toxicants and environmental factors is essential for interpreting toxicity test results, predicting real-world impacts, and refining risk assessments to reflect environmental conditions.

6. Modes of Action: Understanding the modes of action of toxicants on fish at the molecular and physiological levels provides insights into the mechanisms underlying acute toxicity. Toxicants can disrupt vital biological processes, such as respiration, neurological function, and cellular metabolism in fish. Knowledge of these modes of action is essential for understanding the specific biological targets and pathways affected by toxicants and for predicting potential sublethal effects.

7. Risk Assessment and Management: The concepts and principles of acute toxicity inform risk assessment and management strategies aimed at protecting fish populations and aquatic ecosystems. These principles guide the evaluation of potential risks posed by toxicants, the development of risk management options, and the establishment of protective measures and regulatory standards to prevent or mitigate acute toxicity in aquatic environments.

8. Regulatory Standards and Guidelines: Principles of acute

toxicity underpin the development of regulatory standards and guidelines for water quality, chemical management, and environmental protection. These standards are based on acute toxicity data and are designed to safeguard aquatic organisms, including fish, from harmful effects of toxic pollutants. Regulatory measures are essential for ensuring compliance with environmental laws, promoting responsible chemical use, and minimizing adverse impacts on aquatic ecosystems.

Comprehending the concept and principles of acute toxicity of chemicals on fish is vital for protecting aquatic ecosystems, ensuring human health, and promoting sustainable management of water resources. By applying these principles, environmental scientists, regulators, and industry stakeholders can effectively assess and manage the acute toxic effects of chemicals on fish, ultimately contributing to the conservation and resilience of aquatic environments.

Principles

Principles of toxicity of chemicals to fish cover many important aspects, including chemical testing methods, lethality, sub-lethal effects, effective control and development. The principles of acute toxicity of chemicals on fish encompass a range of fundamental concepts that are crucial to understanding the immediate and severe impacts of toxic substances on fish and other aquatic organisms. These principles provide a framework for assessing, managing, and mitigating the acute toxic effects of chemicals in aquatic ecosystems. Key principles of acute toxicity of chemicals on fish include:

1. Bioaccumulation and Bio-magnification: Understanding the principles of bioaccumulation and bio-magnification is essential in assessing the acute toxicity of chemicals on fish. Some toxic substances have the capability to accumulate in the tissues of fish, leading to elevated concentrations within the organism. This bioaccumulation can result in acute toxicity or chronic effects over time. Furthermore, certain chemicals may bio-magnify as they move up the food chain, posing acute toxic risks to fish at higher trophic levels.

2. Behavioral Responses: The examination of behavioral responses forms a fundamental principle in evaluating acute toxicity in fish. Exposure to toxic chemicals can lead to alterations in fish behavior, such as changes in swimming patterns, feeding habits, or avoidance of contaminated areas. Monitoring and understanding these behavioral responses are crucial for early detection of acute toxicity and for assessing the impacts of toxicants on fish populations in their natural habitats.

3. Dissolved and Particulate Toxicants: The distinction between dissolved and particulate toxicants is important in understanding acute toxicity. While dissolved toxicants are freely available in the water column and can directly affect fish through gill respiration and other pathways, particulate toxicants can be ingested by fish through feeding activities. Recognizing the differential impacts of these forms of toxicants is critical for evaluating acute toxicity and managing exposure risks for fish in aquatic environments.

4. Interactions with Other Stressors: The principle of interactions with other stressors acknowledges that fish in natural ecosystems are often exposed to multiple stressors concurrently, such as habitat degradation, temperature variations, and predation pressure. Understanding how these stressors interact with acute toxicants is crucial for assessing the overall impact on fish health and populations. Synergistic or antagonistic interactions between stressors can significantly influence the susceptibility of fish to acute toxicity.

5. Sub-lethal Effects: Acknowledging the occurrence of sub-lethal effects is an important principle in acute toxicity assessments. While acute toxicity tests primarily focus on lethal endpoints, understanding the sub-lethal effects of toxicants on fish, such as impaired reproduction, growth abnormalities, and immune system suppression, provides a more comprehensive view of the impacts on fish populations and long-term ecosystem health.

6. Ecological Relevance: The principle of ecological relevance

emphasizes the importance of considering the ecological context when assessing acute toxicity. It involves evaluating the potential impacts of toxicants on fish within the broader ecological network, including interactions with other species, ecosystem functions, and services. Understanding the ecological relevance of acute toxicity is crucial for making informed decisions regarding chemical management and environmental protection.

7. Early Life Stages: Recognizing the vulnerability of fish early life stages to acute toxicity is a

fundamental principle. Embryos, larvae, and juvenile fish may exhibit increased sensitivity to toxicants, and exposure during these critical stages can have long-lasting effects on fish populations. Assessing acute toxicity in early life stages is essential for understanding the overall impacts of toxicants on fish reproduction and recruitment dynamics.

By incorporating these additional principles into acute toxicity assessments, researchers, regulators, and environmental managers can gain a more comprehensive understanding of the impacts of chemicals on fish and make informed decisions to protect aquatic ecosystems and fish populations.

Important of principles of acute toxicity of chemicals on fish

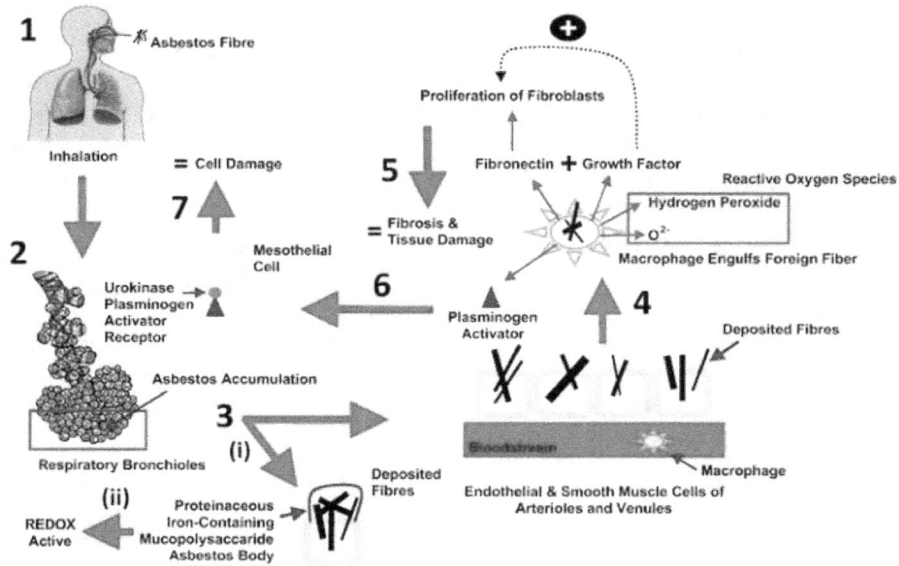

The principles of acute toxicity of chemicals on fish are of utmost importance due to several key reasons:

1. Environmental Protection: Understanding the acute toxicity

of chemicals on fish is crucial for safeguarding the health and integrity of aquatic ecosystems. Fish serve as vital indicators of environmental quality, and acute toxicity assessments help in identifying and mitigating potential risks posed by toxicants, thereby contributing to the protection and conservation of aquatic habitats.

2. Regulatory Standards and Guidelines: The principles of acute toxicity provide the scientific basis for the development of regulatory standards and guidelines aimed at

protecting aquatic organisms. Accurate knowledge of acute toxicity assists regulatory authorities in establishing permissible levels of toxicants in water bodies and formulating effective management strategies to minimize adverse impacts on fish and their habitats.

3. Risk Assessment and Management: These principles play a pivotal role in the assessment and management of risks associated with chemical exposures in aquatic environments. By understanding the potential acute effects of

toxicants on fish, environmental managers can develop risk management plans to prevent, minimize, or mitigate the impact of chemical contaminants on fish populations and aquatic ecosystems.

4. Conservation of Fish Populations: Acute toxicity principles contribute to the conservation and sustainable management of fish populations. By identifying the acute toxicity levels of chemicals, conservation efforts can be targeted to protect vulnerable species, mitigate threats to fish health, and maintain

the ecological balance of aquatic ecosystems.

5. Sustainable Fisheries and Aquaculture: The principles of acute toxicity of chemicals on fish are vital for ensuring the sustainable management of fisheries and aquaculture activities. By understanding the potential impacts of chemical contaminants on fish, measures can be implemented to minimize risks to commercially important fish species and promote responsible practices in the fishing and aquaculture industries.

6. Human Health Protection: Understanding acute toxicity principles is essential for protecting human health, as fish are a significant source of nutrition for many populations. By safeguarding fish populations from toxicant exposures, the potential for adverse effects on human health through consumption of contaminated fish is minimized.

7. Early Detection of Environmental Contamination: Acute toxicity principles facilitate the early detection of environmental contamination

events. By monitoring acute toxicity levels in fish, environmental managers can detect and respond to chemical spills, pollution incidents, or other sources of contamination before widespread ecological damage occurs.

8. Comprehensive Risk Assessment: These principles enable a more comprehensive assessment of chemical risks in aquatic environments. By considering factors such as bioaccumulation, behavioral responses, and interactions with other stressors, a holistic

understanding of the potential acute effects of chemicals on fish and aquatic ecosystems can be obtained.

In summary, the principles of acute toxicity of chemicals on fish are essential for environmental protection, regulatory decision-making, risk management, conservation efforts, sustainable resource management, human health protection, early contamination detection, and comprehensive risk assessment in aquatic environments. By integrating these principles into scientific research, policy development, and environmental management practices, the health and resilience of fish populations and aquatic ecosystems can be effectively safeguarded.

Chapter Three
Practical Applications

The practical application of understanding the acute toxicity of chemicals on fish encompasses various key areas, including environmental protection, regulatory compliance, risk assessment, and conservation efforts. Here are some practical applications of acute toxicity principles:

1. Environmental Monitoring: Acute toxicity testing on fish provides a practical means of monitoring the health of aquatic ecosystems. By assessing the lethal concentration (LC50) of chemicals, environmental managers can detect and quantify the presence of toxicants in water bodies, allowing for early intervention and mitigation

measures to protect fish populations.

2. Regulatory Compliance and Standards: Acute toxicity data on fish serve as a basis for formulating regulatory standards and guidelines to limit the discharge of toxic chemicals into water bodies. This practical application ensures compliance with environmental regulations and promotes responsible chemical use to minimize adverse impacts on fish and aquatic habitats.

3. Risk Assessment for Industrial Processes: Industries that handle potentially hazardous chemicals can apply acute toxicity principles to assess the potential risks to aquatic environments. By conducting acute toxicity tests on fish species, these industries can evaluate the potential impact of accidental releases or ongoing chemical usage, leading to the implementation of risk management strategies to prevent and respond to contamination incidents.

4. Pollution Incident Response: In the event of a chemical spill or

pollution incident, understanding the acute toxicity of chemicals on fish enables rapid risk assessment and appropriate response measures. This may include emergency interventions to minimize harm to fish populations and the wider aquatic ecosystem, as well as the implementation of remediation efforts to restore water quality.

5. Conservation and Management of Aquatic Resources: Acute toxicity data assists in the conservation and sustainable management of fish populations. By identifying potential threats

from chemical contaminants, resource managers can implement measures to protect vulnerable species, preserve biodiversity, and sustainably manage fisheries and aquaculture activities.

6. Development of Safer Chemicals: Understanding the acute toxicity of chemicals on fish provides valuable insights for designing and developing safer chemical compounds. Practical applications include the assessment of chemical alternatives, the modification of chemical formulations to reduce toxicity, and the development of

environmentally friendly products to minimize adverse effects on fish and aquatic ecosystems.

7. Environmental Impact Assessment: In the planning and development of infrastructure projects, such as construction and industrial facilities, the assessment of potential acute toxicity effects on fish is crucial. This practical application allows project proponents to evaluate and mitigate potential impacts on aquatic environments, ensuring compliance with environmental regulations and minimizing harm to fish populations.

8. Fish Health Monitoring in Aquaculture: Acute toxicity principles are applied in aquaculture settings to monitor and maintain the health of farmed fish. By understanding the potential acute effects of waterborne chemicals, aquaculturists can implement measures to ensure optimal water quality, minimize stress on fish stocks, and prevent adverse effects on aquaculture production.

9. Research and Development: Acute toxicity studies on fish contribute to ongoing research and

development efforts in environmental toxicology. Practical applications include the identification of emerging contaminants, the study of chemical interactions, and the advancement of methodologies for assessing acute toxicity, thereby enhancing scientific understanding and informing environmental management practices.

10. Educational and Public Awareness: Understanding the practical implications of acute toxicity of chemicals on fish is essential for educating and raising public awareness about the

importance of protecting aquatic environments. This includes outreach programs, educational initiatives, and public engagement to promote environmentally responsible behaviors and advocate for the conservation of fish populations and aquatic ecosystems.

In summary, the practical applications of understanding the acute toxicity of chemicals on fish are diverse and far-reaching, encompassing environmental monitoring, regulatory compliance, risk assessment, conservation efforts, sustainable resource management, pollution response, development of safer chemicals, environmental impact assessment, aquaculture management, research and development, as well as educational and public awareness initiatives. By integrating these practical applications into environmental management strategies and decision-making processes, the health and resilience of fish populations and aquatic ecosystems can be effectively safeguarded.

Guidelines For Application

Applying the principles of acute toxicity of chemicals on fish, it is crucial to adhere to established guidelines to ensure scientific rigor, ethical considerations, and practical relevance. Guidelines for the application of these principles help to standardize testing methodologies, risk assessment procedures, and environmental management practices. Here are key guidelines for the application of principles of acute toxicity of chemicals on fish:

1. Ethical Considerations: Adherence to ethical guidelines is fundamental when conducting

acute toxicity tests on fish. Researchers and environmental managers must consider the welfare of the test organisms, minimize unnecessary suffering, and ensure that testing procedures comply with ethical standards and regulations governing the use of animals in scientific research.

2. Standardized Testing Protocols: Utilize standardized testing protocols, such as those established by regulatory agencies or international organizations, to conduct acute toxicity tests on fish. Following recognized guidelines ensures consistency,

comparability, and reliability of test results, thereby facilitating sound risk assessment and decision-making.

3. Selection of Test Organisms: Guidelines for the application of acute toxicity principles emphasize the importance of carefully selecting appropriate fish species for testing based on ecological relevance, sensitivity to chemicals, and practical considerations. Consider utilizing standard model species (e.g., rainbow trout, fathead minnow) commonly used in acute toxicity

testing, or species representative of local aquatic ecosystems.

4. Methodological Consistency: Ensure methodological consistency in testing procedures, including environmental conditions, test duration, chemical exposure concentrations, and endpoints for assessing acute toxicity effects on fish. Adhering to standardized methodologies promotes reproducibility and reliability of test results, facilitating meaningful comparisons and risk assessments.

5. Quality Assurance and Quality Control: Implement quality assurance and quality control measures to ensure the accuracy and reliability of test results. Guidelines for quality assurance encompass proper documentation, calibration of equipment, validation of test systems, and adherence to good laboratory practices to minimize sources of variability and error.

6. Data Interpretation and Reporting: Guidelines emphasize transparent and accurate data interpretation, analysis, and reporting of acute toxicity test

results. Ensure that test findings are clearly presented, including statistical analyses, dose-response relationships, and relevant endpoints, enabling informed decision-making and regulatory compliance.

7. Environmental Relevance: Emphasize the environmental relevance of acute toxicity testing by considering factors such as water chemistry, temperature, and the potential interaction of chemicals with environmental matrices. Guidelines stress the need to align testing conditions with the characteristics of the

target aquatic environments to improve the ecological realism of test outcomes.

8. Risk Assessment Framework: Consider established risk assessment frameworks and guidelines when applying acute toxicity principles to evaluate the potential risks posed by chemicals to fish populations and aquatic ecosystems. This involves integrating acute toxicity data with exposure assessment, hazard characterization, and risk characterization to inform risk management strategies.

9. Interdisciplinary Collaboration: Encourage interdisciplinary collaboration among toxicologists, ecologists, environmental scientists, and regulatory experts to ensure a comprehensive and holistic application of acute toxicity principles. Collaboration facilitates the integration of scientific knowledge, regulatory requirements, and practical considerations in addressing acute toxicity issues in aquatic environments.

10. Regulatory Compliance: Adhere to relevant regulatory guidelines and requirements

governing the application of acute toxicity principles, encompassing the use of chemicals, environmental protection, and risk assessment. Compliance with regulatory standards ensures that testing and assessment activities meet legal obligations and contribute to effective environmental management.

11. Continuous Improvement and Adaptation: Embrace a culture of continuous improvement and adaptation by staying abreast of advancements in acute toxicity testing methodologies, scientific understanding, and regulatory

frameworks. Guidelines promote the integration of new knowledge and best practices to enhance the application of acute toxicity principles in environmental management.

By following these guidelines for the application of principles of acute toxicity of chemicals on fish, researchers, environmental managers, and regulatory authorities can ensure the robustness, reliability, and practical relevance of acute toxicity testing, risk assessment, and decision-making processes. Furthermore, adherence to established guidelines promotes the ethical treatment of test organisms, fosters scientific integrity, and supports effective conservation and management of fish populations and aquatic ecosystems.

Case of Acute Toxicity of chemicals on fish

The acute toxicity of chemicals on fish poses a pervasive and often devastating threat to aquatic ecosystems worldwide. As burgeoning industrialization and urbanization escalate the release of hazardous substances into waterways, the urgent need to understand and address acute toxicity challenges has never been more pressing. However, through concerted scientific endeavors, environmental stewardship, and innovative solutions, communities and conservationists are forging a path towards safeguarding fish populations and preserving the integrity of aquatic habitats.

The Peril of Acute Toxicity

The acute toxicity of chemicals on fish manifests as a critical environmental concern, with the potential to imperil the health and survival of aquatic species. Exposure to high concentrations of toxic substances, such as heavy metals, pesticides, and industrial chemicals, can swiftly induce adverse effects in fish, encompassing mortality, physiological stress, impaired reproduction, and behavioral alterations. Moreover, acute toxicity incidents can precipitate far-reaching ecological repercussions, disrupting food webs, compromising water quality, and threatening the resilience of entire aquatic ecosystems.

Understanding the Principles

A fundamental prerequisite in mitigating acute toxicity lies in comprehending the underlying principles governing the interactions between chemicals and fish. The dose-response relationship elucidates the correlation between the concentration of a toxic substance and its impact on fish, crucial for determining lethal concentrations (LC50) and evaluating the potential harm posed by contaminants. Acute toxicity testing, which involves the exposure of fish to varying concentrations of chemicals, offers invaluable insights into species sensitivity, informing risk assessment and management strategies.

Harnessing Multifaceted Strategies

Addressing acute toxicity necessitates a multifaceted approach that encompasses scientific research, environmental monitoring, regulatory frameworks, and community engagement. Rigorous environmental monitoring programs are pivotal in detecting and mitigating acute toxicity incidents, allowing for swift responses to chemical spills and pollution events. Through the establishment of regulatory standards and guidelines, as well as risk assessment protocols, policymakers can mitigate acute toxicity risks and promote the responsible use and disposal of chemicals.

Innovative Solutions
In the relentless pursuit of solutions, innovative technologies and nature-based approaches emerge as powerful tools in combating acute toxicity. Bioremediation, for instance, leverages the natural capabilities of microorganisms and plants to

degrade and detoxify contaminants in aquatic environments, fostering the remediation of polluted habitats. Furthermore, advancements in water treatment technologies and the development of eco-friendly alternatives to hazardous chemicals are instrumental in curbing acute toxicity risks at their source, advocating for sustainable practices and mitigating the environmental footprint of human activities.

Community Engagement and Education

Crucially, the engagement of communities and the dissemination of knowledge play a pivotal role in fortifying the resilience of aquatic ecosystems against acute toxicity. Empowering local stakeholders with environmental education and awareness initiatives fosters a culture of stewardship, instilling a profound appreciation for the intrinsic connection between human actions and the well-being of fish and aquatic environments. Citizen science initiatives further augment environmental monitoring efforts, encouraging public participation in the protection of water bodies and the inhabitants they support.

A Testimony of Resilience
Amidst the perils of acute toxicity, stories abound of communities and conservationists channeling their collective determination and expertise to safeguard fish populations and restore the vitality of aquatic ecosystems. Efforts such as habitat restoration, pollution remediation projects, and sustainable resource management initiatives epitomize the unwavering commitment to minimizing acute toxicity risks and nurturing the health of aquatic

environments. Each success story stands as a testament to the potency of collaboration, innovation, and environmental stewardship in surmounting acute toxicity challenges.

Looking Ahead

As the specter of acute toxicity looms over aquatic ecosystems, the imperative to fortify our defenses and cultivate sustainable coexistence with fish grows ever more urgent. By advancing scientific understanding, harnessing innovative solutions, and fostering community engagement, we can forge a future where fish thrive in vibrant, unperturbed habitats. As custodians of the planet's aquatic riches, it falls upon us to embrace the mantle of conservation and scientific stewardship, safeguarding the invaluable diversity and resilience of fish populations and the ecosystems they inhabit.

Chapter Four
Common Challenges and How
to Overcome Them

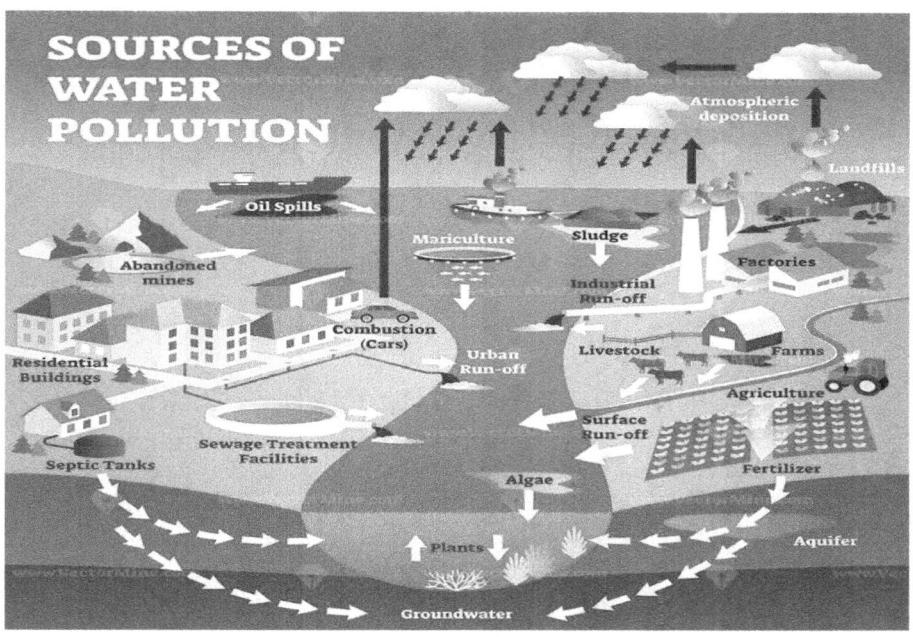

When discussing issues related to
the toxicity of chemicals in fish, it
is important to consider the many
challenges that scientists and

policymakers face in this area. These issues can include limited data and variability in test results, as well as ethical issues and administrative issues. By understanding and resolving these issues, the scientific community can work to improve toxicity assessment of fish and create more efficient and effective tests. Toxicity tests on fish face many challenges that can affect the accuracy and reliability of results. One of the most important problems is that information on the toxicity of fish is limited. This can lead to incomplete or incorrect evaluation of chemicals that have the potential to harm the water system. Additionally, it is difficult to determine a similar and reliable toxicity due to variation in test results due to factors such as temperature, pH and the presence of other chemicals in the water.

Common Challenges

The acute toxicity of chemicals on fish presents a range of complex and interrelated challenges that have significant implications for aquatic ecosystems and the environment as a whole. Some common challenges associated with acute toxicity of chemicals on fish include:

1. Understanding Dose-Response Relationships: Determining the precise relationship between the dosage of a toxic substance and its impact on fish is crucial for assessing the potential risks and developing effective mitigation strategies. However, establishing accurate dose-response relationships can be challenging due to variations in species sensitivity and environmental factors.

2. Species Sensitivity and Variability: Fish species exhibit varying levels of sensitivity to

different chemicals, making it challenging to establish universal standards for acute toxicity across diverse aquatic environments. Understanding and accommodating this species variability is crucial for effective risk assessment and management.

3. Environmental Factors and Interactions: The influence of environmental factors, such as temperature, pH, and water chemistry, on the acute toxicity of chemicals on fish introduces complexities into the assessment of risks and impacts. The interactions between toxicants and

environmental conditions can significantly alter their effects on fish populations, necessitating a comprehensive understanding of these dynamics.

4. Mode of Action and Toxic Pathways: Elucidating the underlying modes of action and toxic pathways of different chemicals on fish is essential for accurately predicting and mitigating acute toxicity. However, the intricate and varied mechanisms through which toxicants affect physiological processes in fish can pose significant challenges for

researchers and environmental managers.

5. Regulatory Standards and Guidelines: Developing and enforcing regulatory standards and guidelines for acute toxicity testing and risk assessment requires collaboration between scientific communities, regulatory bodies, and industry stakeholders. Achieving consensus on standardized testing protocols and risk evaluation frameworks is essential to ensure comprehensive protection for fish populations and aquatic environments.

6. Availability of Data and Research: Access to comprehensive data on the acute toxicity of chemicals on fish, including toxicity thresholds, exposure scenarios, and long-term effects, is essential for informed decision-making and regulatory measures. However, gaps in research and data availability can hinder the development of robust risk assessment and management strategies.

7. Prevention and Mitigation Strategies: Implementing effective measures to prevent acute toxicity incidents and mitigate their

impacts on fish populations requires a multifaceted and integrated approach. Identifying and deploying appropriate prevention and mitigation strategies tailored to specific chemicals and environmental contexts can be a significant challenge.

8. Public Awareness and Stakeholder Engagement: Generating public awareness and fostering stakeholder engagement in addressing acute toxicity challenges is crucial for advocating environmental conservation and sustainable

practices. However, communicating complex scientific information effectively and engaging diverse stakeholders can pose significant communication and outreach challenges.

9. Long-Term Monitoring and Adaptation: Establishing mechanisms for long-term monitoring of acute toxicity impacts and adapting management strategies based on evolving scientific knowledge are critical for ensuring continued protection of fish populations. However, sustained monitoring efforts and adaptive management can be

resource-intensive and require ongoing commitment.

10. Emerging Contaminants and Unknown Risks: The introduction of new and emerging contaminants, as well as the potential synergistic effects of multiple chemicals, present ongoing challenges in anticipating and addressing acute toxicity risks. Understanding and managing the risks posed by emerging contaminants and unknown interactions demand proactive and adaptive approaches.

Solutions to Acute Toxicity of Chemicals on Fish

Addressing the acute toxicity of chemicals on fish requires a multifaceted approach integrating scientific research, regulatory measures, and proactive management strategies. Some key solutions to mitigate this challenge include:

1. Advanced Testing Protocols: Developing and implementing advanced acute toxicity testing protocols that account for species-specific sensitivities, environmental factors, and long-term effects. These protocols

should incorporate the latest scientific knowledge to enhance the accuracy and reliability of toxicity assessments.

2. Integrated Risk Assessment: Adopting an integrated risk assessment approach that considers the interactions between different chemicals, environmental stressors, and the specific physiological responses of fish species. This approach helps in comprehensively evaluating the risks posed by acute toxicity and guiding targeted management strategies.

3. Encouraging Safer Chemical Use: Promoting the development and use of environmentally friendly and less toxic chemicals in industrial processes, agriculture, and consumer products to reduce the potential for acute toxicity in aquatic ecosystems. This can be achieved through incentives, regulatory frameworks, and public awareness campaigns.

4. Environmental Monitoring Networks: Establishing robust environmental monitoring networks to track chemical contamination in aquatic environments, assess acute

toxicity levels, and detect potential pollution incidents. This enables early intervention and informed decision-making to mitigate acute toxicity impacts on fish.

5. Public Awareness and Education: Engaging in outreach programs to raise public awareness about the impacts of chemical pollution on fish and aquatic ecosystems. Educating stakeholders about the importance of responsible chemical use, pollution prevention, and sustainable resource management fosters a culture of environmental stewardship.

6. Regulatory Standards and Compliance: Developing and enforcing stringent regulatory standards for chemical discharges, effluents, and waste disposal to limit acute toxicity risks to fish and other aquatic organisms. Effective enforcement of regulations ensures compliance with established environmental guidelines.

7. Research and Innovation: Supporting research initiatives aimed at understanding the mechanisms of acute toxicity, identifying emerging contaminants, and developing

innovative mitigation strategies. Encouraging collaboration between academia, industry, and governmental agencies can spur advancements in chemical risk assessment and management.

8. Adaptive Management Practices: Embracing adaptive management practices that allow for flexibility in responding to changing acute toxicity threats. This involves ongoing monitoring, assessment of management strategies, and adjustment of approaches based on new scientific findings and environmental conditions.

9. Integrated Water Quality Management: Implementing integrated water quality management approaches that consider the cumulative impacts of various pollutants and stressors on fish health. Managing water quality holistically can help address the root causes of acute toxicity and promote overall ecosystem health.

10. Stakeholder Collaboration: Fostering collaboration among governmental agencies, industry stakeholders, environmental organizations, and local communities to collectively

address acute toxicity challenges. Engaging diverse stakeholders in decision-making processes and management initiatives enhances the effectiveness of mitigation efforts.

By integrating these solutions, stakeholders can work together to mitigate the acute toxicity of chemicals on fish and protect the health and integrity of aquatic ecosystems. This collaborative and proactive approach is essential for promoting sustainable use of chemicals, preserving biodiversity, and ensuring the long-term health of aquatic environments.

Chapter Five

Future Development of Fish Toxicity

In the future, the field of addressing acute toxicity of chemicals on fish is expected to witness significant advancements and innovations aimed at enhancing assessment methods, prevention strategies, regulatory frameworks, and environmental management practices. Key future developments in this area may include:

1. Advanced Toxicity Testing Methods: Continued advancements in toxicity testing methods, including the development of alternative testing models such as in vitro assays, organ-on-a-chip technology, and high-throughput screening approaches. These methods offer rapid, cost-effective, and ethically

preferable alternatives to traditional whole-organism testing, providing valuable data for assessing chemical toxicity on fish.

2. Integrated Omics Approach: Integration of omics technologies such as genomics, transcript omics, proteomics, and metabolomics to better understand the molecular mechanisms underlying acute toxicity in fish. This approach enables comprehensive assessments of the impacts of chemical exposure on biological pathways and the

identification of early biomarkers of toxicity.

3. Predictive Modeling and Big Data Analytics: Utilization of predictive modeling and big data analytics to forecast acute toxicity risks, assess cumulative impacts of multiple stressors, and simulate the effects of chemical exposure under various environmental scenarios. Data-driven approaches can enhance risk prediction, inform decision-making, and optimize resource allocation for mitigation efforts.

4. Nanotechnology and Green Chemistry: Advancements in nanotechnology and green chemistry to develop safer, more sustainable chemicals and materials, reducing the potential for acute toxicity in aquatic ecosystems. Innovations in Nano-enabled products and environmentally benign chemical design can contribute to minimizing adverse impacts on fish health and overall environmental quality.

5. Real-Time Monitoring Technologies: Implementation of real-time monitoring technologies,

such as sensor networks, unmanned aerial vehicles (UAVs), satellite-based monitoring, and autonomous underwater vehicles (AUVs), to continuously track chemical contaminants and acute toxicity levels in aquatic environments. This real-time data can facilitate rapid intervention and decision-making in response to acute toxicity incidents.

6. Precision Environmental Management: Adoption of precision environmental management approaches leveraging remote sensing, geographic information systems

(GIS), and spatial modeling to target acute toxicity hotspots, prioritize conservation efforts, and optimize resource allocation for pollution prevention and remediation activities.

7. Multi-Stakeholder Partnerships: Strengthening multi-stakeholder partnerships and collaborative initiatives involving governmental agencies, industry, academia, non-governmental organizations (NGOs), and local communities to address acute toxicity challenges through shared expertise, resources, and coordinated action.

8. Regulatory Innovations: Evolution of regulatory frameworks to incorporate emerging scientific knowledge, adaptive management principles, and proactive risk assessment strategies. This may involve incorporating updated toxicity testing requirements, expanding the scope of chemical evaluations, and fostering international cooperation to harmonize regulatory standards.

9. Public Engagement and Citizen Science: Expanding public engagement and citizen science initiatives to involve local

communities, citizen scientists, and environmental advocacy groups in monitoring acute toxicity impacts, raising awareness, and contributing to data collection efforts through community-based monitoring programs.

10. Ecosystem-Based Approaches: Embracing ecosystem-based management approaches that consider the interconnectedness of aquatic environments, biodiversity conservation, and sustainable resource use to address acute toxicity challenges comprehensively.

The implementation of these future developments will contribute to enhancing the scientific understanding of acute toxicity of chemicals on fish, improving risk assessment methodologies, fostering the development of safer chemicals, and advancing regulatory and management practices. Furthermore, embracing a holistic and innovative approach will be essential for effectively addressing acute toxicity challenges, safeguarding fish populations, and promoting the long-term health and resilience of aquatic ecosystems.

Future potential Opportunities

The field of acute toxicity of chemicals on fish presents several potential opportunities for future advancements, innovations, and interdisciplinary collaboration. These opportunities can pave the way for addressing acute toxicity challenges and enhancing environmental sustainability. Some potential opportunities include:

1. Advancements in Toxicology and Pharmacology: Opportunities exist to deepen our understanding of the mechanisms of acute toxicity in fish at the molecular and cellular levels. Research in toxicology and pharmacology can lead to the development of targeted antidotes, treatments, and mitigation strategies for chemical

exposures, thereby improving the prognosis for affected fish populations.

2. Integration of Big Data and Artificial Intelligence (AI): The utilization of big data analytics and AI presents opportunities for predicting acute toxicity risks, identifying temporal and spatial patterns of toxic events, and developing early warning systems for monitoring chemical stressors in aquatic ecosystems.

3. Green Chemistry and Sustainable Product Development: Embracing green chemistry

principles and sustainable product development can lead to the creation of eco-friendly alternatives to toxic chemicals, reducing the risks of acute toxicity in fish while promoting environmental sustainability.

4. Innovative Monitoring Technologies: The development and deployment of innovative monitoring technologies, including sensor networks, unmanned aerial vehicles (UAVs), satellite imaging, and autonomous underwater vehicles (AUVs), offer opportunities to enhance real-time detection and monitoring of toxic

substances in aquatic environments.

5. Public-Private Partnerships for Research and Development: Collaborative initiatives involving government, industry, academic institutions, and non-governmental organizations can facilitate the pooling of resources, expertise, and technology to drive research and development efforts aimed at mitigating acute chemical toxicity in fish.

6. Community-Based Monitoring and Citizen Science: Engaging local communities and citizen

scientists in monitoring and reporting acute toxicity events can offer valuable opportunities for collecting data, raising awareness, and fostering a sense of shared responsibility for protecting fish and aquatic ecosystems.

7. Regulatory Harmonization and Standardization: Opportunities exist to promote international collaboration and harmonization of regulations and standards related to chemical safety and acute toxicity testing, ensuring a consistent approach to risk assessment and management across different regions.

8. Education and Public Awareness Campaigns: Educational programs and public awareness campaigns can play a crucial role in enhancing understanding of the risks associated with chemical toxicity in fish, promoting responsible chemical use, and encouraging informed decision-making among stakeholders.

9. Funding for Research and Innovation: Increased investment in research funding and innovation initiatives can accelerate progress in understanding acute toxicity

mechanisms, developing effective mitigation strategies, and advancing the development of less toxic alternatives to hazardous chemicals.

10. Interdisciplinary Collaboration and Knowledge Exchange: Encouraging interdisciplinary collaboration between scientists, policymakers, environmental engineers, chemists, biologists, and other stakeholders can foster knowledge exchange and the development of comprehensive solutions to acute toxicity challenges.

By capitalizing on these potential opportunities, the field of acute toxicity of chemicals on fish can make significant strides in mitigating risks, protecting aquatic ecosystems, and promoting the sustainable coexistence of human activities and environmental health. Embracing innovation, collaboration, and a commitment to sustainable practices will be key in unlocking the potential opportunities to address acute toxicity challenges effectively.

Future Predictions and insights

Predicting the future of acute toxicity of chemicals on fish involves recognizing emerging trends, technological advancements, regulatory developments, and environmental shifts that may impact the field. Here are some future predictions and insights on acute toxicity of chemicals on fish:

1. Integration of Omics Technologies: The incorporation of omics technologies, including genomics, transcript omics, proteomics, and metabolomics, will enable a more comprehensive

understanding of the molecular responses of fish to chemical exposures. This will enhance our ability to predict acute toxicity outcomes and develop targeted interventions.

2. Expansion of Predictive Modeling: Advances in computational toxicology and predictive modeling will allow for the simulation of acute toxicity scenarios, considering multiple factors such as chemical properties, environmental conditions, and species-specific sensitivities. This will aid in risk assessment and management.

3. Focus on Mixtures and Synergistic Effects: The recognition of the combined effects of chemical mixtures and the potential synergistic interactions between compounds will drive research towards elucidating the complexities of acute toxicity in real-world multi-contaminant scenarios.

4. Emphasis on Early Life Stages: Greater attention will be given to understanding the susceptibility of early life stages of fish to chemical toxicity, considering the potential

long-term impacts on population dynamics and ecosystem health.

5. Increased Use of Bioassay Technologies: Bioassay techniques, such as in vitro assays and high-throughput screening platforms, will gain prominence for rapidly assessing acute toxicity and identifying hazardous substances, facilitating more efficient risk evaluations.

6. Adoption of Alternative Testing Methods: The adoption of alternative testing methods, including in silica approaches, organ-on-a-chip technologies, and

3D cell cultures, will reduce reliance on traditional animal testing and offer more ethically and environmentally sustainable means of evaluating chemical toxicity.

7. Innovations in Monitoring and Detection: The development of novel sensors, autonomous monitoring systems, and remote sensing technologies will enable real-time detection and monitoring of acute toxicity events, enhancing environmental surveillance capabilities.

8. Regulatory Evolution: Regulatory frameworks will evolve to address emerging contaminants, refine risk assessment methodologies, and incorporate new scientific findings to enhance the protection of fish and aquatic environments from acute chemical toxicity.

9. Emphasis on Circular Economy and Green Chemistry: Increased focus on circular economy principles and the adoption of green chemistry practices will drive the development of safer chemicals and sustainable manufacturing processes, aiming

to minimize the occurrence of acute toxicity incidents.

10. Public Engagement and Citizen Science: The active involvement of citizens in monitoring aquatic environments, reporting pollution incidents, and participating in citizen science initiatives will contribute to a more comprehensive understanding of acute toxicity occurrences and support community-driven conservation efforts.

11. Global Collaboration and Knowledge Sharing: Enhanced

international collaboration, knowledge exchange, and capacity-building initiatives will promote cross-disciplinary cooperation, aiding in the development of common frameworks for addressing acute toxicity challenges on a global scale.

These future predictions and insights underscore the need for proactive measures to address acute toxicity of chemicals on fish, involving advancements in science, technology, regulation, and societal engagement. By leveraging these developments, the field can anticipate and respond to emerging challenges while advancing the protection and conservation of fish and aquatic ecosystems.

Chapter Six
Summary

Acute toxicity of chemicals on fish is a critical concern with significant implications for aquatic ecosystems, human health, and regulatory frameworks. Understanding the principles and factors involved in acute toxicity is essential for effective risk assessment, management, and conservation efforts. Key aspects include dose-response relationships, species sensitivity, environmental interactions, modes of action, and regulatory standards. Practical applications encompass environmental monitoring, regulatory compliance, risk assessment, pollution response, and conservation initiatives. Anticipating the future of acute toxicity involves advancements in omics technologies, predictive modeling, bioassay methods, regulatory evolution, and public engagement, with a focus on sustainable practices and global collaboration. These insights highlight the complex interplay between scientific, regulatory, and societal factors in addressing acute toxicity challenges and safeguarding fish and aquatic environments.

Recommendations For Action

Addressing acute toxicity of chemicals on fish requires a multifaceted approach involving various stakeholders. Here are some recommendations for action:

1. Enhanced Monitoring and Surveillance: Implement comprehensive monitoring programs to track chemical contamination in aquatic environments, focusing on areas of high vulnerability and sensitive fish habitats. Utilize advanced

technologies for real-time detection and early warning systems to promptly respond to acute toxicity incidents.

2. Research and Development: Encourage and invest in research initiatives to better understand the mechanisms of acute toxicity, species-specific responses, and the effects of chemical mixtures. This could involve exploring alternative testing methods, omics technologies, and interdisciplinary collaborations to advance scientific knowledge.

3. Regulatory Frameworks: Work towards harmonizing and strengthening regulations related to chemical use, discharge, and environmental exposure. Incorporate advances in toxicity testing methods, risk assessment approaches, and emerging contaminant considerations into regulatory standards to ensure fish protection and ecosystem health.

4. Industry Practices: Promote the adoption of green chemistry principles, sustainable manufacturing processes, and responsible chemical management practices within industrial sectors.

Encourage the development and use of safer alternatives to reduce the risk of acute toxicity to fish.

5. Public Awareness and Education: Conduct outreach efforts to raise awareness about the impacts of chemical toxicity on fish and aquatic ecosystems. Engage the public in citizen science initiatives, pollution reporting, and conservation actions to empower communities to contribute to environmental protection efforts.

6. International Collaboration: Foster collaboration at the global

level to share best practices, data, and technologies for addressing acute toxicity challenges on a broader scale. Engage in knowledge exchange, capacity building, and joint research initiatives to develop common frameworks for fish protection.

7. Early Life Stage Protection: Emphasize the protection of early life stages of fish through targeted conservation measures, regulation of developmental toxicants, and inclusion of early life stage testing in toxicity assessments to safeguard populations and biodiversity.

8. Sustainable Development Goals Integration: Align actions to address acute toxicity with broader sustainable development goals, such as those related to clean water and life below water, to ensure comprehensive and holistic approaches to environmental protection.

9. Adaptive Management: Implement adaptive management strategies to respond to evolving acute toxicity challenges, incorporating new scientific findings, emerging contaminants, and changing environmental

conditions into management and regulatory frameworks.

10. Collaboration Across Sectors: Foster collaboration between government agencies, academic institutions, industry stakeholders, environmental organizations, and local communities to develop integrated approaches and share expertise in addressing acute toxicity challenges.

By implementing these recommendations, stakeholders can work collectively to mitigate the risks associated with acute toxicity of chemicals on fish, protect aquatic ecosystems, and ensure the long-term health and resilience of fish populations.

Take Action Now

Addressing the acute toxicity of chemicals on fish requires a comprehensive approach involving multiple actions aimed at monitoring, prevention, and mitigation. Here are some recommended actions to take:

1. Environmental Monitoring: Establish comprehensive monitoring programs to regularly assess water bodies and aquatic environments for the presence of

toxic chemicals. This includes monitoring sites near industrial facilities, agricultural areas, and other potential sources of chemical contamination.

2. Research and Risk Assessment: Invest in research to better understand the toxicity of chemicals on fish, including their effects on different species, life stages, and ecosystems. Conduct thorough risk assessments to identify high-risk areas and vulnerable fish populations.

3. Regulatory Measures: Strengthen and enforce regulations

related to the use, discharge, and management of toxic chemicals. Implement strict controls on the release of pollutants into water bodies and establish stringent water quality standards to protect fish and aquatic life.

4. Pollution Prevention: Promote pollution prevention measures in industrial, agricultural, and urban settings to minimize the introduction of toxic chemicals into aquatic systems. Encourage the use of environmentally friendly practices and the adoption of less toxic alternatives.

5. Public Awareness and Education: Conduct outreach programs to raise public awareness about the impacts of chemical toxicity on fish and the environment. Empower communities to report pollution incidents and participate in conservation efforts to protect local waterways and fish populations.

6. Emergency Response Planning: Develop and regularly update emergency response plans to address acute toxicity incidents in water bodies. Establish protocols for rapid response, containment,

and cleanup of chemical spills to minimize the impact on fish and other aquatic organisms.

7. Sustainable Practices: Encourage the adoption of sustainable practices in industries that use or produce chemicals. Promote the principles of green chemistry, waste reduction, and the safe handling and disposal of toxic substances to protect aquatic ecosystems.

8. Habitat Protection: Implement measures to preserve and restore critical fish habitats, such as wetlands, spawning areas, and

nursery grounds. Protecting these habitats contributes to the overall resilience of fish populations in the face of chemical threats.

9. Collaboration and Partnerships: Foster collaboration between government agencies, research institutions, industry stakeholders, environmental organizations, and local communities. Collective efforts can lead to the development of effective solutions and the sharing of best practices.

10. Long-Term Monitoring and Evaluation: Establish long-term monitoring programs to track

changes in fish populations, species diversity, and ecosystem health in response to efforts to mitigate acute toxicity. Regular evaluation of these measures can guide future interventions.

By taking these actions, stakeholders can work towards reducing the acute toxicity of chemicals on fish, protecting aquatic ecosystems, and ensuring the sustainability of fish populations for future generations.